IMAGES of America
NORFOLK NAVAL SHIPYARD

On the Cover: Pres. Franklin D. Roosevelt arrives to inspect work in progress at the shipyard on July 29, 1940. USS *Potomac* (AG-25), the presidential yacht, is in the background. Roosevelt used his "Floating White House" until his death in 1945. It is now a museum in Oakland, California. (Courtesy of the Navy History and Heritage Command.)

IMAGES of America
NORFOLK NAVAL SHIPYARD

Lt. Christopher Miller, USN (Ret.)

ARCADIA
PUBLISHING

Copyright © 2018 by Lt. Christopher Miller, USN (Ret.)
ISBN 978-1-4671-2976-3

Published by Arcadia Publishing
Charleston, South Carolina

Printed in the United States of America

Library of Congress Control Number: 2018937809

For all general information, please contact Arcadia Publishing:
Telephone 843-853-2070
Fax 843-853-0044
E-mail sales@arcadiapublishing.com
For customer service and orders:
Toll-Free 1-888-313-2665

Visit us on the Internet at www.arcadiapublishing.com

This book is dedicated to the men and women of Norfolk Naval Shipyard, past and present, who have devoted their lives to building, repairing, and maintaining America's navy.

CONTENTS

Acknowledgments		6
Introduction		7
Norfolk Naval Shipyard Timeline		8
1.	The Yard	9
2.	The People	37
3.	The Ships	61

Acknowledgments

The majority of the digital materials in this book come from the archives of the Naval History and Heritage Command (NHHC). Unless otherwise noted, all images appear courtesy of the Naval History and Heritage Command. The NHHC compiled the materials because of generous donations by the Norfolk Naval Shipyard, Portsmouth Naval Shipyard Museum, and the Hampton Roads Naval Museum. The Library of Congress and Boston Public Library were also the source of many photographs and postcards. Furthermore, I would like to express my appreciation to the staff of the Public Affairs Office at the Norfolk Naval Shipyard. Their expertise and dedication were critical to telling the story of the shipyard. I thank them all for continuing to preserve the history of the US Navy. I would also like to take the opportunity to thank the University of Phoenix for encouraging me to work on this book. Its commitment to employee professional development is inspiring. Additionally, the staff of Arcadia Publishing, especially my editor, Angel Hisnanick, made writing my first book such a pleasure. I can't wait to do it again. Finally, I would like to thank my husband, James, for his support while I was writing this book. I could not have done it without him. In the end, hundreds of thousands of men and women have dedicated their lives to working at and supporting Norfolk Naval Shipyard. Consequently, quite a few stories wait to be told about the shipyard's contributions to our navy and our nation. This is one sailor's perspective.

INTRODUCTION

Norfolk Naval Shipyard (NNSY) was founded in 1767 by Andrew Sprowle, a local merchant. It was initially named Gosport to tie it to an area of the same name in the South Hampshire area of England; it was also previously known as the Norfolk Navy Yard. It is the oldest shipyard owned and operated by the US Navy. Prior to the Revolutionary War, the shipyard was quite prosperous as a naval repair and commerce facility. Because of its ideal location on the shore of the Elizabeth River, British troops burned it to the ground in 1779. After leasing the land from Virginia in 1794, the new American navy built USS *Chesapeake* (one of the now-famous six frigates) from the keel up in 1799. During the next 60 years, the shipyard expanded its footprint and capability.

In 1861, Virginia seceded from the Union, so the commander of the yard ordered it burned to prevent its use by Confederate forces. Again, the location and capabilities of the shipyard made it too valuable to fall into enemy hands. The strategy did not work. Confederate forces took over the shipyard and a valuable stockpile of war-making material. CSS *Virginia* (known by many as USS *Merrimack*) was built in 1862, shortly before the Confederacy burned the shipyard again prior to its capture by the Union. The shipyard was renamed Norfolk Navy Yard in 1862. To this day, the name is controversial across Hampton Roads, because the shipyard is physically located in Portsmouth. However, the naming made sense because, at the time, it was located in Norfolk County. Additionally, the US Navy already had a shipyard named Portsmouth in Maine.

After the Civil War, the shipyard continued to be a vital asset to the Navy with respect to shipbuilding, repair, resupply, and, interestingly, homeporting. The yard was instrumental in refitting and resupplying ships during the Spanish-American War in 1898. In 1917, Naval Station Norfolk was founded, and the ships that were homeported at the shipyard were transferred up the river. The shipyard was expanded around this time to accommodate employees and their families, because shipbuilding and repair continued. In fact, some of the most famous ships of the 20th century were either built or modernized at NNSY. For example, USS *Arizona* received its modernization at the yard in the early 1930s. During World War II, the shipyard underwent another expansion to keep up with the needs of the naval station. During the war years of 1939–1945, the shipyard constructed 101 ships of all types and provided repairs to 6,850 naval vessels of both the American and Allied fleets. After the war, the yard suspended shipbuilding operations with the final new construction in 1953 and shifted primarily to repair and overhaul of every type of vessel the US Navy operates, including nuclear power. Today, it focuses solely on nuclear-power vessels.

Norfolk Naval Shipyard remains the oldest shipyard owned and operated by the US Navy. Take a walk across the Jordan Bridge (which connects South Norfolk and Portsmouth), and you can see aircraft carriers and submarines covered with scaffolding and shipyard workers. They, combined with crewmembers, keep the Atlantic Fleet prepared to meet any and all adversaries to the United States.

Norfolk Naval Shipyard Timeline

Four flags have flown over Norfolk Naval Shipyard: those of Great Britain, Virginia, the United States, and the Confederacy.

It has been burned three times—in 1779, 1861, and 1862—but never stopped operating since its founding 250 years ago.

It is the oldest of the Navy's four shipyards.

At its peak during World War II, it employed nearly 43,000 people, built 101 vessels of all kinds, and repaired 6,850 others. Below are some key dates in its history:

1752   The town of Portsmouth is established by Col. William Crawford.

1767   November 1, Gosport Shipyard is established in Portsmouth by Andrew Sprowle, a prominent merchant and British loyalist.

1775   April 19, the battles at Lexington and Concord kick off the Revolutionary War.

   October 13, the US Navy is born when the Continental Congress votes in Philadelphia to outfit two warships.

   December 9, troops under Lord Dunmore, Virginia's last royal governor, are defeated at the Battle of Great Bridge.

1776   January 1, Dunmore leaves Gosport, where he had been headquartered. His ships bombard Norfolk, where a cannonball lodges in the wall of St. Paul's Episcopal Church.

1779   May 15, a British fleet of six warships, under Commodore Sir George Collier, burns the Gosport Shipyard and burns or captures 137 vessels in the harbor.

1781   October 19, Lord Cornwallis surrenders to American and French forces at Yorktown, effectively ending the Revolutionary War.

1784   Portsmouth annexes Gosport lands, though the state of Virginia keeps ownership of the shipyard.

1794   Congress passes the "Act to Provide a Naval Armament," authorizing the building of six frigates and the loaning of the Gosport yard to the United States by Virginia.

1798	April 30, the US Department of the Navy is established, and the yard becomes known as Gosport Navy Yard.
1800	May 22, the frigate USS *Chesapeake*, built at Gosport, is commissioned.
1801	June 15, the federal government buys the 16-acre shipyard from Virginia for $12,000.
1821	A school for midshipmen is established at Gosport, 24 years before the founding of the US Naval Academy.
1829	July 11, Pres. Andrew Jackson visits Gosport, where one of the nation's first two dry docks is under construction.
1830	Portsmouth Naval Hospital opens and becomes an annex of the shipyard.
1833	June 17, Gosport's new dry dock opens for business, receiving USS *Delaware*, the first ship dry docked in America.
1846	Commanding officer Jesse Wilkinson uses his own money to buy 40 acres across the Elizabeth River from Gosport, creating St. Helena Annex. He later sells the land to the government.
1861	April 12, Confederates bombard Fort Sumter in Charleston, South Carolina. Five days later, Virginia votes to secede from the Union. April 20–21, Union forces abandon and burn Gosport Navy Yard, as well as 11 ships, including the steam frigate USS *Merrimack*. July, Confederate shipyard workers begin converting the unburned underbelly of USS *Merrimack* into the ironclad CSS *Virginia* in Drydock 1.
1862	March 8–9, in the Battle of Hampton Roads, CSS *Virginia*, fresh from Drydock 1, destroys USS *Cumberland* and USS *Congress*, killing 337 sailors, before battling the ironclad USS *Monitor* to a draw. May 10, Confederate forces burn and abandon the Gosport yard; the shipyard is renamed US Navy Yard Norfolk. May 11, CSS *Virginia* is scuttled off Craney Island before dawn, hours before USS *Monitor* accompanies Pres. Abraham Lincoln, aboard USS *Baltimore*, on the Elizabeth River to the shipyard to survey fire damage.
1876	Sunken CSS *Virginia* is removed as a navigation hazard off Craney Island and taken to Drydock 1, its birthplace, where it is broken apart.
1892	USS *Texas*, the nation's first battleship, built at Gosport, is launched.
1901	USS *Holland*, the nation's first submarine, arrives at the shipyard for evaluation and repairs.
1904	Dr. William Schmoele Jr. sells the government 273 acres of open space along the Elizabeth River, vastly expanding the shipyard's footprint.

1907	December 16, Pres. Theodore Roosevelt reviews an armada of 16 battleships—the "Great White Fleet"—as it leaves Hampton Roads on a round-the-world tour to project US naval power. The shipyard helped prepare three of the ships for the journey.
1917	April 6, the United States enters World War I. October 12, Naval Station Norfolk, the world's largest naval base, opens.
1918	Construction starts on two planned communities to house shipyard workers: Truxtun for African Americans and Cradock for whites
1921	USS *Langley*, the nation's first aircraft carrier, completes its conversion from the collier *Jupiter* and sets sail.
1929	February 13, the yard is designated Norfolk Navy Yard.
1931	USS *Arizona* completes its three-year modernization, 10 years before it is sunk at Pearl Harbor.
1940	July 23, work is completed on the Hammerhead Crane.
1941	December 8, the United States declares war on Japan; on December 11, it declares war on Germany and Italy.
1945	December 1, yard is renamed Norfolk Naval Shipyard.
1953	USS *Bold* and USS *Bulwark* become the last ships built at the shipyard.
1965	The shipyard attains nuclear technology capability when USS *Skate* (SSN-578) becomes the first nuclear submarine to undergo a major overhaul.
2001	The Hammerhead Crane ceases operations.
2012	The aircraft carrier USS *George H.W. Bush* (CVN-77) arrives at the shipyard for the first time for a planned incremental availability.
2015	USS *La Jolla* (SSN-701) arrives at the shipyard for conversion to a moored training ship (MTS).
2017	The shipyard successfully undocks USS *Rhode Island* (SSBN-740) two days early and completes the USS *Harry S. Truman* (CVN-75) planned incremental availability two days early. Norfolk Naval Shipyard celebrates its 250th anniversary.

One

THE YARD

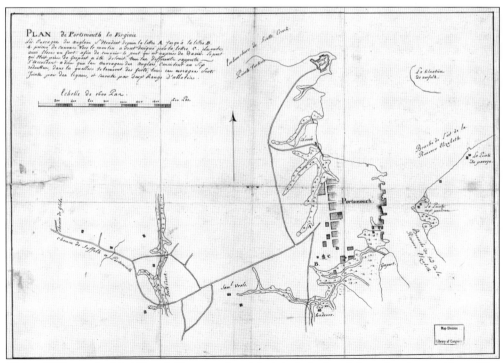

This c. 1791 map is titled *Plan de Portsmouth en Virginie*. It depicts defense barriers and forts. The Gosport Shipyard is depicted in the bottom right corner.

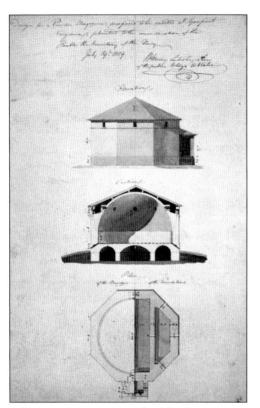

This architectural drawing for a powder magazine is dated July 19, 1809. Powder magazines such as this one were essential. Gunpowder was stored here in wood barrels.

A sketch map of the US Navy yard vicinity in Portsmouth shows buildings, shipbuilding docks, a dry dock, buoys, and the locations of the scuttled US ships *Pennsylvania*, *Columbia*, *Raritan*, and *Delaware*.

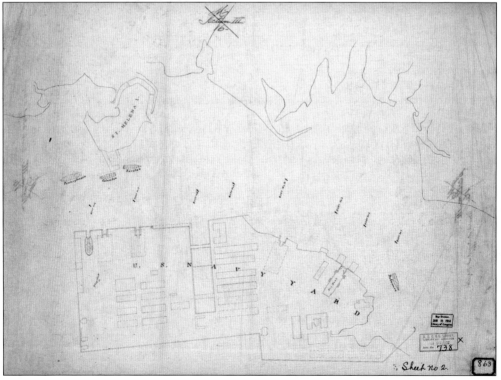

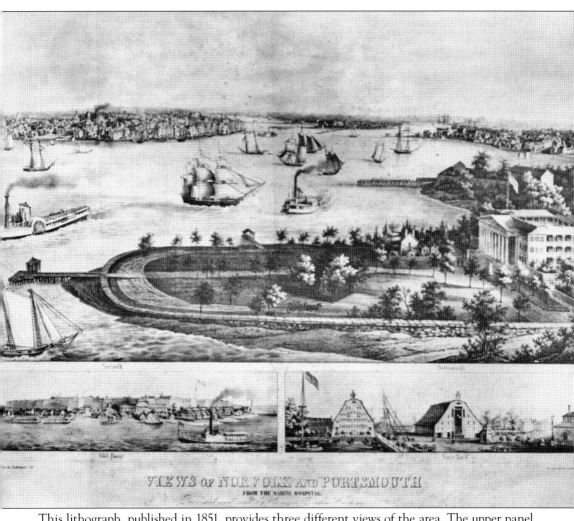

This lithograph, published in 1851, provides three different views of the area. The upper panel shows the naval hospital in the foreground with the city of Portsmouth behind it. The large building on the right with the columns still stands today. The Portsmouth Naval Hospital is a short boat ride down the Elizabeth River to the Navy Yard. In the bottom right panel is Old Point Comfort, which would become a critical chokepoint in the Civil War 10 years later. The lower left panel shows the Navy Yard as it was in 1851.

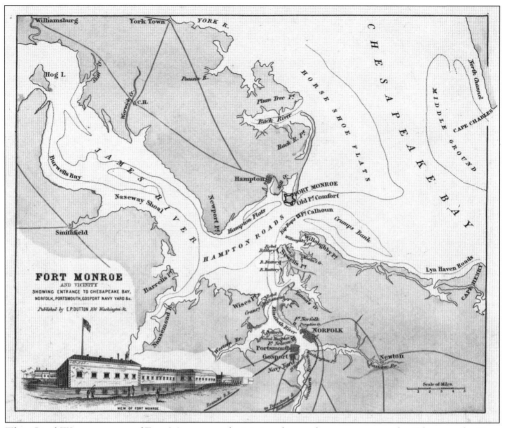

This Civil War–era map of Fort Monroe and vicinity shows the entrance to the Chesapeake Bay, Norfolk, Portsmouth, and the Gosport Navy Yard. The lighthouse at Fort Monroe is still used as a primary navigation aid for ships entering and leaving Norfolk.

This small tar-and-pitch facility produced materials for wood preservation. Cooking tar resulted in caustic fumes, hence the need for a standalone building.

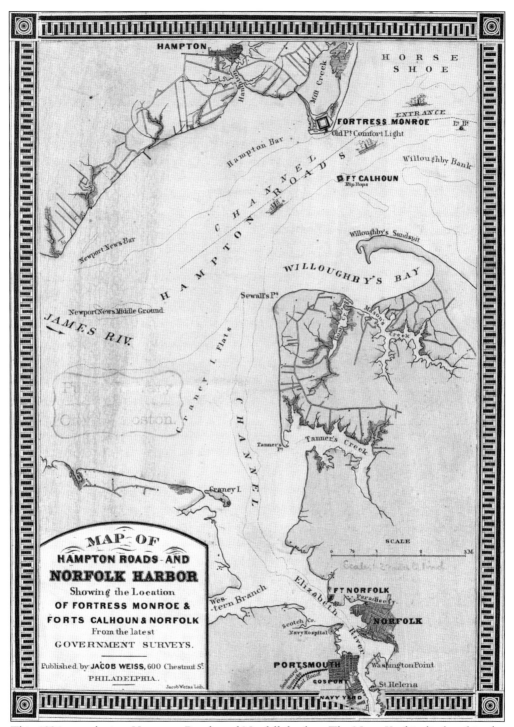

This 1894 map depicts Hampton Roads and Norfolk harbor. The Navy Yard is displayed in the bottom right corner of the map.

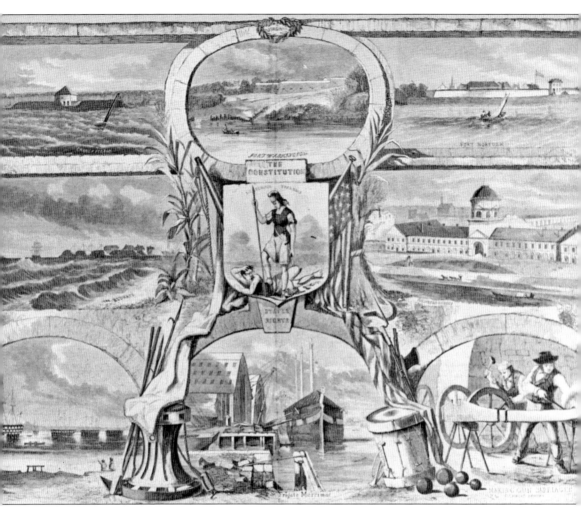

A line engraving published in *Harper's Weekly* in 1861 provides a series of individually labeled Virginia scenes and symbols on the eve of the Civil War. Clockwise around the edge of the print from the upper right are Fort Norfolk; the Richmond Armory; gun carriages being made at the Richmond Armory; USS *Merrimack* at Norfolk Navy Yard; inactive warships *Pennsylvania*, *Columbia*, *Raritan*, and *United States* at Norfolk Navy Yard; fortifications on the riprags at Hampton Roads; and Craney Island. Fort Washington, in Maryland across the Potomac River from Virginia, is shown at top. In the center is a representation of the Virginia state seal and motto (*Sic Semper Tyrannis*), with "The Constitution" and "States' Rights" inscribed above and below it.

Outside of the yard's west wall, the remains of rifle embrasures used by Confederate defenders are seen in 1861–1862. Rifles embrasures are essentially holes cut in walls that provide a wide field of fire with minimal risk to the shooter.

This line engraving published in Harper's Weekly in 1861 provides two scenes of the burning of Norfolk Navy Yard and the destruction of several ships by the Union. Ships shown in the lower scene (as identified below the print), from left to right, are USS *United States* (burning); tug *Yankee* with USS *Cumberland* (underway, leaving the area); USS *Merrimack* (burning in the left center distance); USS *Pawnee* (underway, leaving the area), and USS *Pennsylvania* (burning).

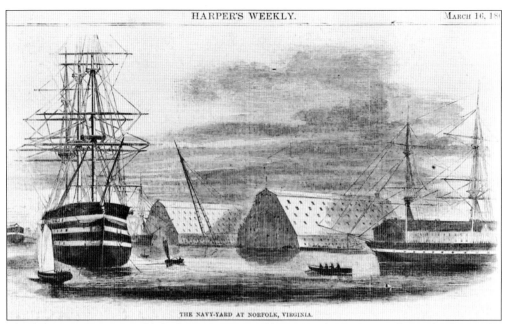

This line engraving was published in *Harper's Weekly* on March 16, 1861. It shows the ship of the line *Pennsylvania*, the receiving ship at the yard, at anchor on the left side of the image. The Navy Yard and *Pennsylvania* were burned a little more than a month later, on April 20, 1861.

This pitch house was one of the Navy Yard's oldest landmarks until it was torn down in 1943 to make room for waterfront improvements. It stood near Building 28 at South Landing.

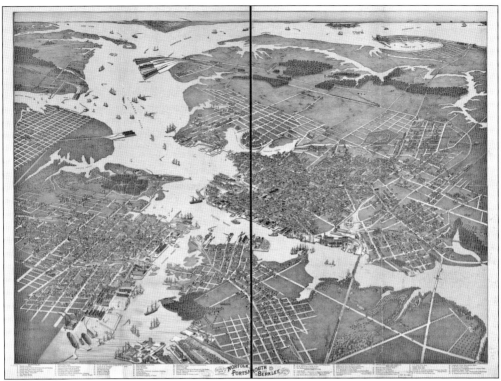
This detailed map depicts a bird's-eye view of Norfolk, Portsmouth, and Berkley in 1891. The Navy Yard is labeled in the bottom left corner.

These submarine chasers, shown under construction in 1917, were built specifically to hunt down German submarines during World War I. They were relatively small (about 80 feet) and could travel around 20 knots. Sub chasers like these were outfitted with guns and, more importantly, paravanes, which were towed cables with attached explosives that would detonate when contacted by a submarine. Building 8 and a large crane are in the background.

This waterfront scene is from the autumn of 1907. Ships at left are (from front to rear) USS *Hopkins* (DD-6), USS *Lawrence* (DD-8), USS *Hull* (DD-7), USS *Talbot* (TB-15), and USS *Moccasin* (SS-5). The latter two are hauled out on the marine railway. USS *Stewart* (DD-13) is in the right foreground. Ahead of it are a torpedo boat, a barge, and the tug *Mohawk*.

This photograph depicts the opening ceremonies for Drydocks 6 and 7 in 1919. The honorary guests were the king and queen of Belgium.

This aerial photograph was taken by a US Army Air Corps plane on June 1, 1921. USS *Langley* (CV-1), the first American aircraft carrier, is in the left background. It was converted from USS *Jupiter* (AC-3). In addition to its groundbreaking role as aircraft carrier, it was also the first turboelectric-powered ship. Other ships present include the newly commissioned USS *Dahlgren* (DD-187) and USS *Goldsborough* (DD-188).

This photograph was taken at the 100th anniversary of Drydock 1 at the Norfolk Navy Yard in 1933. The caption to this step memorial reads "Opened 17 June 1833. Andrew Jackson, President of the United States. Levi Woodbury, Secretary of the Navy. Loammi Baldwin, Engineer. One Hundredth Anniversary. 17 June 1933. Franklin D. Roosevelt, President of the United States. Claude A. Swanson, Secretary of the Navy."

This aerial photograph depicts St. Julien's Creek Annex (SJCA), a facility next to the shipyard that began operations as a naval facility in 1849. Past operations at SJCA have included general ordnance operations involving wartime transfer of ammunitions to various other US naval facilities throughout the United States and abroad. In 1898, the facility was equipped for assembling ammunition, and until 1970, it supplied ammunition to the fleet in addition to loading, assembling, issuing, and receiving naval gun ammunition and conducting experimental and test loading for new ammunition.

The Mast House served as the Norfolk Naval Shipyard's mast construction facility from 1828 to 1904, during the height of sailing vessels. The building had to be quite large to facilitate the manufacture of the large mainsails of ships like USS *Chesapeake*, whose mainsail was almost 90 feet wide at the bottom.

Another view of the Mast House shows the large size of the building.

This photograph depicts the pumping out of Drydock 8 for the first time in August 1941. The forms that retained the concrete underwater and the remaining portion of the back creek not yet filled in can be seen.

A waterfront view of the shipyard shows the receiving ship *Franklin* in the distance. Once ships like *Franklin* were no longer useful for navigation and combat, they were often transformed into floating barracks.

This photograph depicts Dry Dock No. 4, a 1,000-foot dry dock under construction around 1918. It was then the largest dry dock in the world, and it was constructed almost entirely of concrete.

The Alexander Park civilian worker housing project is pictured under construction in the fall of 1942. The expansion of the shipyard to assist with the war effort facilitated the need for affordable housing for shipyard workers.

This photograph depicts Pier 5 under construction in November 1938. Pilings are being driven at the outboard end, and excavation is underway at right. Drydock 8 was built in the creek in the background later.

This photograph depicts the rebuilding of Drydock 2 during the mid-1930s. Public Works Administration funds in 1933 provided for the replacement of the old timbered side walls with concrete.

This postcard depicts the USO Club on Crawford and South Streets in Portsmouth. It was a popular hangout for shipyard workers. (Courtesy of the Boston Public Library.)

This postcard depicts the main gate of the Norfolk Navy Yard. It is likely from the late 1930s based on the vehicles and was taken by Clipper Studios. (Courtesy of the Boston Public Library.)

A dry dock caisson is launched in 1933 for the Norfolk Navy Yard. A caisson is essentially a large gate at the water end of a dry dock. It is flooded with water to seat it in place.

This photograph depicts one of the dry docks at the Norfolk Navy Yard. Based on vehicles in the far background, this photograph was likely taken in the 1920s. The sign in the foreground states, "Make no lines to this caisson."

"The Hammerhead Crane" at Norfolk Naval Shipyard in Portsmouth, Virginia, completed in 1940, is pictured in this 1970 oil painting, which measures 24 by 30 inches. USS *Rankin* (KA-103) is pierside.

This picture shows the Great Steam Crane in 1905. This crane served from the 1890s to the 1940s and was vital to daily shipyard operations. (Courtesy of the Library of Congress.)

The up-and-down bridge at Norfolk Naval Shipyard is shown in this 1970 oil painting, which measures 24 by 30 inches. The bridge is up in the painting to accommodate the outbound warship.

Building 173 burns to the ground on August 4, 1944. The building contained the commissioning outfit for USS *Shangri-La* (CV-38).

A fire destroys Pier 3 at the St. Helena Annex on January 16, 1945. YFB-43 and curious spectators, including a couple of sailors, are at left.

This rendering shows the extension to Building 163 at the Navy Yard. It was commissioned by the Navy Department's Bureau of Yards and Docks. The extension to the Structural Shop is today known as Building 234, the Sheetmetal Shop.

This photograph depicts several tugboats on the left and a larger warship in the center. The ships are unidentified. This image was taken from the roof of Building 60, the power plant in 1908.

USS *Abel P. Upshur* (DD-193) is pictured in 1921. It was a *Clemson*-class destroyer that served the Navy until it was transferred to the Royal Navy in 1940.

This overhead photograph of Norfolk Naval Shipyard was taken in 1969. Although not pictured here, the shipyard includes seven dry docks, two shipways, 45 ship berths, a railway system, and nuclear sub overhaul, making it the largest shipyard in the world devoted to naval ship repair and conversion.

This photograph depicts Quarters A, the historic quarters of the shipyard commander. The home was built in the late 1830s in the Greek Revival style. It is the most elaborate of the officers' quarters on the yard.

This photograph depicts Quarters B and C, historic officers' quarters at the yard. These residences, along with Quarters A, were added to the National Register of Historic Places in 1974.

Two

THE PEOPLE

This painting depicts the officers and crew of HMS *Shannon* boarding and capturing USS *Chesapeake* in 1813. *Chesapeake*, one of the original six frigates, was built at Norfolk Navy Yard.

The remains of the Norfolk Navy Yard are pictured after it was burned during the Civil War. A lone man is shown in the photograph looking out toward the Elizabeth River. This December 1864 image depicts Civil War damages to the yard.

The receiving ship USS *Franklin* and several steam launches are depicted here in the early 1900s. The sailors appear very happy to have their photographs taken.

This all-female yeoman drill team was stationed at the Norfolk Navy Yard. The leader of the team, Rosa Ober, holds the drill team flag. An important part of the life of yeomen during this time period was recreation. Primarily, this was athletic in nature and included drill, sailing, and boat racing. The yeoman F rating was authorized by the Naval Reserve Act of 1916. These women, as well as the Nurse Corps, paved the way for the women in today's naval services.

This oversized panoramic photograph shows the female yeomen of the Supply Department at Norfolk Navy Yard. It was taken on March 14, 1919. The yeoman rating was one of the only job

specialties available to women in the early part of the 20th century.

The commanding officer of USS *Camden* (AS-6), Capt. William Miller, proudly poses for a c. 1920 photograph onboard his ship in the Fitting Out Basin. *Camden* was a seized German cargo ship.

Shipyard workers proudly stand beside boilers bound for USS *Craven* in 1917. Craven proudly served the Navy from 1917 to 1922. In 1940, it was transferred to the Royal Navy and renamed HMS *Lewes*. It served throughout World War II.

This photograph depicts USS *Eagle* (PE-17) moored pierside in 1921. It is clear that the sailors knew they were being photographed.

This panoramic photograph of USS *Gulfport* (AK-5) officers and crew posing in front of their ship at the Norfolk Navy Yard was taken on January 13, 1921. The ship's commanding officer, in the center, is probably Lt. Comdr. S.W. Hickey. *Gulfport* was acquired by the Navy in Hawaii at the start of World War I. Prior to its acquisition, it was a privately owned German steamship. During its short five-year service life, it provided critical cargo support between ports on the East Coast.

This 1921 photograph depicts payday onboard USS *Bushnell*. As one might imagine, the depicted sailors are quite jubilant at the prospect of having cash in their pockets.

Secretary of the Navy Edwin Denby (left) greets Rear Adm. Philip Andrews, commandant of the Norfolk Navy Yard, on board USS *Henderson* (AP-1) on May 20, 1922. The ship was about to leave the Norfolk Navy Yard to take the Naval Academy class of 1881 to Japan.

This photograph depicts officers and crew proudly standing beside USS *Borie* (DD-215) in 1927. *Borie* was a *Clemson*-class destroyer that saw action in the Battle of the Atlantic during World War II. It was named for Secretary of the Navy Adolphe Borie, who served under Ulysses S. Grant.

This photograph depicts officers and crew of USS *Bushnell* (AGS-5) standing on and beside their ship in 1921. It was a submarine tender and was named in honor of David Bushnell, who invented the American submarine.

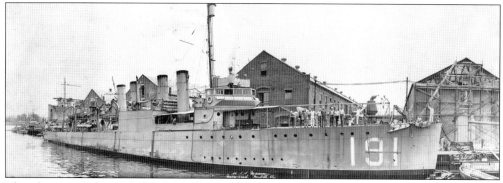
This photograph depicts USS *Mason* (DD-191) moored pierside in 1921 with several crewmembers looking on. Like *Craven*, it was transferred to the Royal Navy in 1940 through a bases for ships exchange program. As HMS *Broadwater*, it proudly patrolled the waters of the North Atlantic until it was sunk by a German U-boat in 1941.

The Navy Yard's Crane Division is depicted in 1921. Cranes, large and small, remain critical to daily operations at the shipyard.

Rear Adm. Felix Gygax, yard commandant, bids "good hunting" to a Navy Yard recruiter bound for West Virginia in a sound truck around 1943–1945. Note the poster on the truck featuring USS *Shangri-La* (CV-38).

The turret lathe section of Shop 31 is pictured here. Turret lathes were critical machining equipment during World War II because of their ability to make duplicate parts. This shop was primarily staffed by women when this photograph was taken in 1943.

This worker is in the plate-bender's shop, where steel plates are bent to fit the sidings of ships under construction. This is an operation requiring a high degree of skill and strength. (Courtesy of the Library of Congress.)

USS *Tucker*'s launching ceremony in 1936 is pictured. Mrs. Leonard P. Thorner, the ship's sponsor, is receiving the commemorative champagne bottle used to welcome the ship to the Navy.

Officers review a nurses' parade held at the Navy Yard's Marine Barracks on October 13, 1943. They are, from left to right, Rear Adm. Felix Gygax, commander, Norfolk Navy Yard; Comdr. Helen M. Bunty, chief nurse, Norfolk Naval Hospital; Capt. G.E. Thomas (Medical Corps), commanding officer, Norfolk Naval Hospital, and Col. J.R. Henley, USMC, commanding officer, Marine Barracks.

Firemen examine debris onboard USS *Saturn* after a fire on board had killed 15 workmen while the ship was undergoing conversion to a refrigerator ship in 1944.

USS *Tarawa* (CV-40) is christened by Mrs. Julian C. Smith on May 12, 1945. *Tarawa* was an *Essex*-class aircraft carrier that served the Navy during the Korean War. It was decommissioned in 1960.

Capt. Elliott Laughlin (left), commanding officer of USS *Mississinewa* (AO-144) accepts the Battle Efficiency plaque from Rear Adm. Ira H. Nunn in the ship's wardroom while the fleet oiler was undergoing a general overhaul at the shipyard in 1957.

This photograph depicts a sailing ship anchor from around 1820. It was on display at the shipyard in the 1940s until it was removed to support a wartime scrap-metal drive.

A panoramic photograph shows USS *Huron* pierside in 1919. *Huron* was formerly SS *Friedrich der Grosse*, a German commercial freighter that was seized by the United States and served the US armed forces as a troop transport throughout World War I.

Nurses parade in review in 1943 on the parade ground with their Marine Corps drill instructors. Similar training still takes place for new officers at Naval Station Newport in Rhode Island.

This photograph depicts officers and crew of USS *Paducah* in 1921. It was a *Dubuque*-class gunboat that served the Navy in a variety of capacities until World War II.

This photograph depicts huge guns being assembled. They will later be installed on one of the many warships being built during World War II at the shipyard. (Courtesy of the Library of Congress.)

The sheet metal department at the shipyard is pictured in 1941. Sheet metal is used for a variety of purposes onboard naval vessels because it is easy to keep clean and is resistant to rust. (Courtesy of the Library of Congress.)

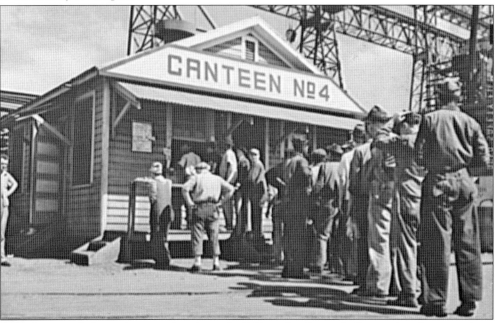

This photograph shows Canteen 4 in the early 1940s at the shipyard. Shipyard workers are taking a break from a hard day's work to stop for lunch. This canteen was one of several at the yard. (Courtesy of the Library of Congress.)

This c. early 1940s photograph depicts the welding shop in operation. Welding remains a critical component of shipbuilding. Much stronger than rivets, welding also allows for easy cuts and repairs. (Courtesy of the Library of Congress.)

A shipyard worker inspects a four-bladed propeller bound for a large warship in this c. early 1940s photograph. These massive propellers are cast in one piece to increase their efficiency in the water. (Courtesy of the Library of Congress.)

This welder is torch-flaming a part for a new destroyer in this c. early 1940s image. Welding parts like this one creates a strong bond to the ship, but it also allows for cutting to replace and repair worn parts of the ship. (Courtesy of the Library of Congress.)

Vice Adm. Patrick N.L. Bellinger, US Navy, congratulates Comdr. Richard Edward III, US Navy Reserve, after presenting the bronze star during the commissioning ceremony for USS *Lake Champlain* (CV-39) in June 1945. A large shipyard crane is visible in the background.

This outstanding panoramic photograph is from the commissioning ceremony for USS *Alabama* (BB-60) in August 1942. Hundreds of sailors in their dress whites stand by to take *Alabama* to sea. This crew would later participate in several notable campaigns during World War II in both the Atlantic and Pacific areas of operation.

Captain Edsall (right) shakes hands with Captain Sylvester (left) after he relieves him as commanding officer of USS *Missouri* (BB-63). This change of command ceremony was held on board the ship at the shipyard in September 1952. Captain Edsall died less than a year later of heart failure aboard *Missouri*. The life ring between the two gentlemen remains to this day onboard the ship in Pearl Harbor, Hawaii.

Crewmembers of USS *Norfolk* (DL-1) are depicted topside during its decommissioning ceremony at the yard in 1970. It served as the flagship for the South Atlantic forces during the 1960s.

Three

THE SHIPS

USS *Chesapeake* (pictured underway here), one of the original six frigates authorized by the Naval Act of 1794, was built at the yard. It has 38 guns and three heavy masts.

In 1833, the Gosport Shipyard (Norfolk) and the Charlestown Shipyard (Boston) were in a race to see who would have the first operational dry dock. Norfolk won when USS *Delaware*, depicted here, entered the dry dock to much fanfare on June 17, 1833.

Just a couple of years after its historic dry docking period at the Norfolk Navy Yard, USS *Delaware* (along with USS *Potomac*) was struck by a white squall in the Gulf of Lyons on October 11, 1835. This painting was donated to the Navy History and Heritage Command by Pres. Franklin D. Roosevelt in 1936.

This line engraving was published in *Harper's Weekly* in 1861 as part of a larger print. It depicts (from left to right) the ships *Pennsylvania, Columbia, Raritan,* and *United States* moored off the yard prior to their destruction on April 20, 1861.

"Virginia" Sinking the "Cumberland," March 8, 1862.

This engraving shows Virginia sinking USS Cumberland at Hampton Roads (a name given to the convergence of the Elizabeth, James, and Nansemond Rivers and the Chesapeake Bay). Virginia remains one of the most famous ships built at the Navy Yard.

This line engraving depicts USS *Merrimack* being repaired in early 1861. It was burned when Union forces abandoned and destroyed the yard on April 20, 1861.

This engraving depicts CSS *Virginia* in dry dock in early 1862. It was then nearing completion after conversion from the burned hulk of USS *Merrimack*.

This engraving depicts CSS *Virginia* at sea near the Norfolk Navy Yard. *Virginia* was the first steam-powered ironclad built by the Confederacy. It was a primary participant in the Battle of Hampton Roads.

Receiving ships USS *Richmond* (left) and USS *Franklin* are moored at the Navy Yard in 1898. These ships, no longer useful for combat, were used to receive new Navy recruits as they awaited assignment to their new ships. As this photograph was taken during the Spanish-American War, recruiting was high to support the predominantly naval war effort.

This photograph depicts the US lighthouse tender *Maple* at the Norfolk Navy Yard in 1898. The diary entry was written by Richard G. Davenport. The entry summarizes the *Maple*'s part in the 1898 blockade of Cuba during the Spanish-American War.

USS *Holland*, the first modern American submarine, was serviced at the shipyard. This would start a long and proud tradition of service to submariners at the yard.

This photograph depicts USS *Nashville* (PG-7) off the Norfolk Navy Yard in 1898. After the sinking of *Maine* in 1898, *Nashville* was assigned to the North Atlantic Fleet. Many of its sailors were awarded the Medal of Honor for their service during the Spanish-American War.

US Navy tug *Alice* is pictured in 1900 moored pierside at the Norfolk Navy Yard. *Alice* was initially built for commercial tug operations, but it was purchased by the Navy in 1898 during the Spanish-American War to serve as a supply tug. It spent the next 18 years providing tug support to naval operations in the area. *Alice* was struck from the register in 1916, one year before the United States entered World War I.

This photograph depicts USS *Annapolis* (PG-10) moored pierside in 1900. The large gunboat provided vital service to the Navy during the Spanish-American War, especially during the blockade of Cuba.

USS *Glacier* (AF-4) is moored pierside at the Norfolk Navy Yard in 1905. *Glacier* served in the Spanish-American War and provided support to the Philippines before being decommissioned in June 1903. Six months later, it was recommissioned in Norfolk and proceeded to provide critical supply support to ships in the deadly North Atlantic during World War I. Interestingly, *Glacier* was decommissioned again and then recommissioned before its final decommissioning in 1922.

This postcard depicts torpedo boats in Wet Slip 1 in 1905. The US Navy experimented heavily with several different types of torpedo boats in the latter half of the 19th century.

This photograph depicts TB-24-class torpedo boats in 1907. They were initially designed as defensive boats. However, the emergence of the more capable destroyer relegated most of these ships to coastal patrol activities and training platforms. This photo postcard was published as a souvenir of the 1907 Jamestown Exposition.

Torpedo boats of the Atlantic Fleet Reserve Torpedo Flotilla are pictured at the Norfolk Navy Yard around 1907. Most of these craft are partially dismantled. The two boats in the front right and the one in the front left (listed in no particular order) are USS *Bagley* (TB-24), USS *Barney* (TB-25), and USS *Biddle* (TB-26). The two larger boats between them in the foreground are USS *DuPont* (TB-7, left) and USS *Porter* (TB-6). In the back row are, from left to right, a torpedo boat (*Foote*, *Rodgers*, or *Winslow*); USS *Cushing* (TB-1), and either USS *Gwin* (TB-16) or USS *Talbot* (TB-15). The receiving ship USS *Franklin* (1867–1915) and a two-masted schooner are in the distance.

This postcard depicts USS *Pennsylvania* (ACR-4) during the 1907 Jamestown Exposition. Two ships of sail and one tender are also pictured.

The FORECASTLE LOG
U. S. S. TENNESSEE
At Sea. - - Norfolk Navy Yard.

FEBRUARY 26, 1916. At sea. At 5:45 a.m. the jack staff was carried away by a heavy sea. *FEB. 27.* Topmast seemed to bend under the heavy wind and so slowed down to 1-3 speed and sent men aloft to put on preventors. All hands on the top side basking in the sunshine, I wonder why? *FEB. 28.* At 1:27 a.m. anchored off the coast of North Carolina. At 6:55 a.m. got underway again. At 7:52 a.m. anchored again. The Tennessee is certainly there when it comes to picking out those long interesting cruises. How about King Neptune. At 12:29 p.m. got underway again at a speed of 11 knots. At 4:25 p.m. anchored in Hampton Roads, Va. Put all the passengers ashore. *FEB. 29.* At 8:23 a.m. got underway for the Norfolk Navy Yard. Fired a 13 gun salute to the Commandant of the Norfolk Navy Yard. Commenced coaling at 1:00 p.m. Knocked off coaling at 6:00 p.m. *MARCH 1.* Commenced coaling again at 6:30 a.m. Finished at 11:30 a.m., having taken on board 1512.5 tons. Liberty party went ashore. *MAR. 2.* Liberty party returned on a port tact. Great place Norfolk. Sent a 48 hour liberty party ashore. Big doings tonight. *MAR. 3.* Very cold all day. Making the ship ready for our trip around South America, putting in the latest style of bed rooms, etc., and taking stores on board. Foolish question No. 000,000,000,00 how much liberty do we get on this trip? Distance traveled from Port-au-Prince, 1,190 miles. Distance from Hampton Roads around South America and return about 14000 miles.

This photograph is of a 1916 USS *Tennessee* deck log entry. Highlights include a 13-gun salute to the commandant of the Navy Yard.

USS *Teaser* (SP-933) is pictured in the Elizabeth River off the Norfolk Navy Yard around 1917–1918. Built as a civilian motorboat in 1916, this craft was placed in commission by the Navy on November 29, 1917. It caught fire and sank on December 27, 1918, and was stricken from the list of naval vessels on February 15, 1919.

USS *Courtney* (SP-375) is pictured off the Norfolk Navy Yard on August 18, 1917. It is painted in pattern camouflage.

This photograph depicts USS *Craven* (DD-70) ready for launching at the yard in 1918. It served the US Navy until it was transferred to the Royal Navy and renamed HMS *Lewes*.

This oversized 1919 panoramic of the USS *Wisconsin* (BB-9) shows the first battleship to bear the name. The second battleship USS *Wisconsin* (BB-64) is now a museum on the Norfolk waterfront.

USS *Noa* (DD-343) is moored pierside at the shipyard in 1921. *Noa* was a *Clemson*-class destroyer that served the US Navy following World War II. At the beginning of World War II, it was converted into a troop transport destroyer.

This photograph depicts USS *Nevada* (BB-36) leaving Norfolk Navy Yard on January 4, 1920. It was the first of two *Nevada*-class battleships and served in both World Wars I and II. *Nevada* was the only ship to get underway when the Japanese attacked Pearl Harbor in 1941.

This panoramic photograph was taken by the G.L. Hall Optical Company in 1919. USS *Madawaska* is in World War I dazzle camouflage, more than three months after the Armistice. The Navy Yard's railway crane is in the right center foreground.

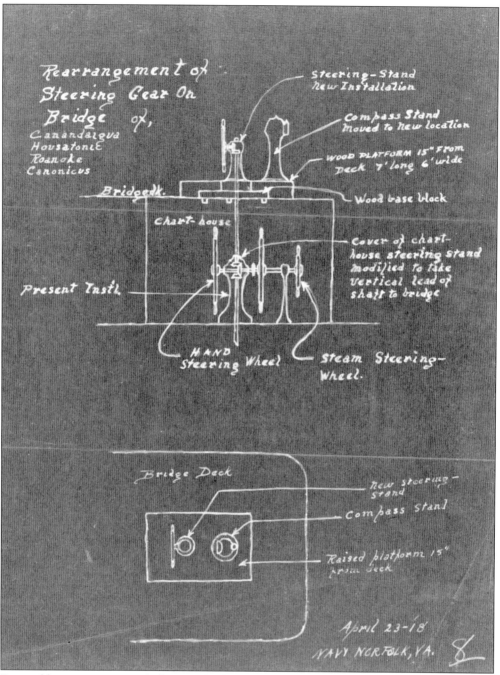

Pictured here is an engineer's drawing for the rearrangement of bridge steering gear. The ships slated for the modification were *Canandaigua*, *Housatonic*, *Roanoke*, and *Canonicus*. This drawing is dated April 23, 1918.

This photograph depicts, from left to right, an unidentified ship, USS *Abel P. Upshur* (DD-193), USS *Welborn C. Wood* (DD-195), and USS *Branch* (DD-197) in 1921, located in the great Wet Slip.

USS *Robin* (AM-3) is in dry dock on March 11, 1921. *Robin* was a minesweeper that operated extensively along the East Coast of the United States. In 1942, it was designated as an ocean-going tug.

This photograph depicts USS *Nereus* (AC-10) moored pierside in 1921. *Nereus* was a *Proteus*-class collier (cargo ship).

This c. 1901 photograph depicts Drydock 2 at the yard. USS *Ajax* is in for repairs.

The minesweeper USS *Auk* (AM-57) is pictured underway near the Navy Yard. It was built in Norfolk and commissioned in 1942.

USS *Arizona* (BB-39) is moored pierside in 1932 after completing modernization. *Arizona* was one of only two *Pennsylvania*-class battleships built during World War I. Although it did not see action during the war, it did provide escort duties to President Wilson while he traveled to the Paris Peace Conference in 1919. Just nine years after this photograph was taken, *Arizona* and over half its crew would lie at the bottom of Pearl Harbor.

This photograph depicts USS *Mississippi* (BB-41) and USS *Idaho* (BB-42) undergoing repairs at the yard in 1933. *Mississippi* and *Idaho* were both *New Mexico*–class battleships. They were both sold for scrap after World War II.

This is the 1937 completion photograph of USS *Tucker* (DD-374). *Tucker* was a *Mahan*-class destroyer. *Tucker* was present in Pearl Harbor during the surprise Japanese attack on December 7, 1941.

USS *New York* (BB-34), at left, and USS *Texas* (BB-35) are pictured at the shipyard in the 1920s. Both ships were *New York*–class battleships and both served during World Wars I and II.

This photograph depicts USS *Oklahoma* (BB-37) pierside awaiting a captain's inspection in 1921. Of particular note in this picture are the exposed secondary guns and men working.

This photograph depicts the production schedule for USS *Alabama* (BB-60) in 1942. It was commissioned on August 16, 1942.

LST-334 is pictured off the Norfolk Navy Yard with LCT-481 on deck around 1943. LST-334 was built at the Norfolk Navy Yard and served the Navy from 1942 until 1946, when it was decommissioned.

USS *South Dakota* (BB-57) is pictured at anchor off the Navy Yard in 1943. It participated in multiple operations during World War II and was later sold for scrap.

This photograph depicts PT-109 in dry dock at the yard. This small boat gained notoriety because of one of its skippers, Pres. John F. Kennedy.

This photograph depicts submarine chaser PC-1594 off the Navy Yard in 1944. The small but fast vessels were designed specifically for antisubmarine warfare.

USS *Texas*, the first American battleship, shown here underway, was commissioned in 1895 at Norfolk Navy Yard. *Texas* was built at Norfolk Navy Yard. The ship saw significant action during the Spanish-American War.

These postcards were published by the Nesson Sales Company of Norfolk from 1930 to 1945. The battleship in dry dock (above) could be USS *Arizona*. The postcard below depicts battleships and a destroyer in dry dock. (Both, courtesy of the Boston Public Library.)

This postcard depicts a battleship entering the harbor near the shipyard. The postcard was published by the Nesson Sales Company of Norfolk between 1930 and 1945. The ship is likely USS *Arizona*. (Courtesy of the Boston Public Library.)

This postcard shows battleships, destroyers, and a large crane. The postcard was published by the Nesson Sales Company of Norfolk between 1930 and 1945. (Courtesy of the Boston Public Library.)

This postcard's main action is an aircraft carrier in wet dock. This is another postcard published by the Nesson Sales Company. (Courtesy of the Boston Public Library.)

This tanker is laid up for minor repairs in 1941. During World War II, the Norfolk Naval Shipyard was quite busy building, maintaining, and repairing vessels for the United States and repairing quite a few for allied nations. (Courtesy of the Library of Congress.)

This photograph depicts LCM landing craft under construction in the early summer of 1942. The Norfolk Navy Yard built a total of 50 LCMs like the ones shown here.

USS *Kentucky*'s (BB-66) keel is prepared for launching on June 10, 1942, to clear the shipway for landing ship tank (LST) construction. Work was not resumed on *Kentucky*'s hull for nearly 30 more months.

This photograph depicts USS *Liberty* (GTR-5) moored pierside at the Navy Yard. In 1967, this technical research ship was attacked by Israel during the Six-Day War.

USS *Alabama* (BB-60) is pictured being launched in 1942. This was the sixth ship named after the state of Alabama. The current ship to bear the name is a ballistic missile submarine.

USS *Iwo Jima* (LPH-2) and *Harlan County* (LST-1196) are pictured together in Drydock 8 in 1977. In front of the lock is USS *Nahoke* (YTM-536). *Iwo Jima* served the Navy from 1960 until 1993. It was the first ship designed solely for amphibious operations. Consequently, it saw action in the Vietnam War, providing troops and supplies (including aircraft) for the war. In 1970, it was part of the now-famous Task Force 170, recovering the command module from Apollo 13's failed mission to the moon. It is not surprising that it is being serviced next to USS *Harlan County*, a *Newport*-class tank landing ship whose mission was amphibious support, primarily cargo.

USS *Skate* (SSN 578) was one of the first nuclear-powered submarines and the first nuclear submarine serviced at the Norfolk Naval Shipyard. She entered the shipyard in 1965 for refueling and overhaul, which included the SUBSAFE package. She was the first submarine to receive the package, which was designed and implemented after the loss of the USS *Thresher* in 1963.

This photograph depicts USS *Massachusetts* laid up at the shipyard in 1963. At that time, it had been stricken from the Naval Vessel Register but was awaiting preservation as a memorial. USS *Uvalde* (AKA-88) is at right. *Massachusetts* is now a floating museum in Fall River, Massachusetts.

USS *Kentucky*'s (BB-66) bow is being transported on a large crane barge from Newport News, Virginia, to the Norfolk Naval Shipyard in May 1956. It was used to repair USS *Wisconsin* (BB-64), which had been damaged in a collision on May 6, 1956. The tug closest to camera is *Alamingo* (YTB-227). The tug on other side of barge is *Apohola* (YTB-502).

This photograph depicts the *Jackson*-class transport ship USS *President Jackson* (APA-18) underway off the Navy Yard in 1947. The ship was critical in providing reinforcements and evacuating casualties during World War II and the Korean War.

This photograph depicts USS *Osprey* in 1941 conducting post-shipyard sea trials. *Osprey* was a *Raven*-class minesweeper that was commissioned in 1940.

This photograph depicts *Gearing*-class destroyer USS *Carpenter* (DD-825) moored pierside in the early 1950s. It served the Navy until 1981, when it was transferred to Turkey. *Carpenter* also makes a cameo in the film *Raise the Titanic*.

USS *Kentucky* (BB-66) is pictured at Newport News Shipbuilding & Drydock Company in 1956. The hull is complete only to the second deck and is missing its bow, which was removed to repair USS *Wisconsin* (BB-64) earlier in the year.

This photograph depicts USS *Hartford* (IX-13) rotting away at the shipyard after World War II. USS *Midway* (CVB-41) and the Hammerhead Crane are in the right background.

USS *Patoka* (AO-9) approaches the Norfolk Navy Yard in 1932. It is being assisted by one of the yard's harbor tugs.

This photograph depicts USS *Blue* (DD-387), at left, and USS *Helm* (DD-388), ready for christening in Drydock 2 in 1937. They were both *Bagley*-class destroyers.

USS *Kentucky* (BB-66) is pictured under construction at the shipyard in 1946. It was never completed and was sold for scrap in 1958.

USS *Osmand Ingram* (AVD-9) is underway off the Norfolk Navy Yard in 1943. It was built as a *Clemson*-class destroyer but was later converted to a seaplane tender. During World War II, *Osmand Ingram* resumed operations as a destroyer, even earning the Presidential Unit Citation.

USS *Patoka* (AO-9) clears the Norfolk Navy Yard in 1932. It was probably photographed by Callahan, a local photography studio.

This photograph depicts USS *Bancroft* in the stone dry dock in the early 1900s. Of particular note in this photograph is the horse and tow in the right background.

USS *Goodrich* (DDR-831) is pictured during an inclining experiment at the shipyard. *Goodrich* was a *Gearing*-class destroyer that visited the Navy Yard for its FRAM II modernization, a program to upgrade World War II–era destroyers to extend their lifespan. The program was started by Adm. Arleigh Burke.

The YD-73 floating crane lowers weights onto the deck of USS *Goodrich* (DDR-831) during an inclining experiment at the shipyard. Inclining is a test performed on ships to determine their stability, weight, and center of gravity.

This photograph depicts the USS *Yorktown* (CV-10) crew standing at attention as the national ensign is raised during the ship's commissioning ceremony at the shipyard in 1943. *Yorktown* was an *Essex*-class carrier that served the Navy in a variety of capacities, including recovering the Apollo 8 command module.

USS *George Washington* (CVN-75) conducts a test of its countermeasure wash down system in 1997 during post-shipyard sea trials. (Courtesy of Naval Sea Systems Command.)

This is an aerial view of the US Navy aircraft carrier USS *America* (CV-66) in dry dock at the Norfolk Naval Shipyard on June 1, 1982. *America* was decommissioned in 1996 after serving the Navy for 30 years.

Crewmembers stand topside as the guided-missile submarine USS *Florida* (SSGN-728) departs Norfolk Naval Shipyard in 2006 and transits through downtown Norfolk toward its new homeport of Naval Submarine Base King's Bay in Georgia. *Florida* entered Norfolk Naval Shipyard to undergo a refueling and conversion from a ballistic missile submarine (SSBN) to the new class of guided missile submarines (SSGN). The nuclear-powered submarine now has the capability to launch up to 154 Tomahawk cruise missiles, conduct sustained special forces operations, and carry other payloads, such as unmanned underwater vehicles (UUVs), unmanned aerial vehicles (UAVs), and special forces equipment. (Courtesy of Naval Sea Systems Command.)

This aerial view shows the Norfolk Naval Shipyard on the Elizabeth River in 1995. At the bottom right is the South Gate Annex, where ships of the mothball fleet are stored. At left center is the main shipyard. The city of Norfolk is in the background. (US Navy photograph.)

The U.S. Navy guided-missile destroyer USS *Ramage* (DDG-61) is pictured in a floating dry dock at the Norfolk Naval Shipyard on May 25, 2012. *Ramage* is the 11th *Arleigh Burke*–class destroyer.

A Norfolk Naval Shipyard crane removes the main mast from the island structure of USS *George Washington* (CVN-73) in 2005. The Norfolk-based *Nimitz*-class aircraft carrier was undergoing a $300-million shipyard availability at Norfolk Naval Shipyard. (US Navy photograph by Mass Communication Specialist Seaman Jennifer Apsey.)

The amphibious assault ship USS *Kearsarge* (LHD-3) commences a dry dock flooding operation at Norfolk Naval Shipyard. Kearsarge was undergoing a 10-month dry dock planned maintenance availability and was scheduled to go underway in the fall of 2018. (US Navy photograph by Mass Communication Specialist First Class Emmitt J. Hawks.)

On August 13, 2003, USS *George Washington* (CVN-73) transits out of the Norfolk Naval Shipyard in Portsmouth to the Atlantic Ocean to conduct sea trials in preparation for a scheduled deployment. (US Navy photograph by Photographer's Mate Third Class Mark Martinez.)

Norfolk Naval Shipyard employee Percy Moton, of Welding Shop 26, patches a bulkhead during general repairs onboard USS *Harry S. Truman* (CVN-75). *Truman* was conducting a Docked Planned Incremental Availability at Norfolk Naval Shipyard and was scheduled to return to sea in the fall of 2018. (US Navy photograph by Mass Communication Specialist Seaman Joshua A. Moore.)

This 2014 photograph depicts USS *Harry S. Truman* (CVN-75) heading to the shipyard. (US Navy photograph by Mass Communication Specialist Seaman Justin R. Pacheco.)

The *Ohio*-class ballistic missile submarine USS *Maryland* (SSBN-738) is pictured after completing its refueling overhaul in 2012. *Maryland*, like its sister SSBNs, is capable of carrying 24 Trident II D5 missiles. This class of submarine is so large that is has nearly the same draft as a modern aircraft carrier. *Maryland* is homeported in King's Bay, Georgia. (US Navy photograph by Tony Anderson, NNSY photographer.)

This 2015 photograph depicts USS *George H.W. Bush* (CVN-77) arriving at the shipyard for a planned incremental availability (repair period). Ships rely on these availabilities for critical upgrades and to extend their service life. (US Navy photograph by Shayne Hensley, NNSY photographer.)

This 2016 photograph depicts aircraft carrier USS *Harry S. Truman* (CVN-75) transiting the Elizabeth River from its homeport at Naval Station Norfolk to Norfolk Naval Shipyard. (US Navy photograph by Mass Communication Specialist Seaman Victoria Granado.)

The aircraft carrier USS *Dwight D. Eisenhower* (CVN-69) is assisted by several tugs in 2016 as it transits the Elizabeth River on its way to the shipyard. *Eisenhower*'s homeport is Naval Station Norfolk. (US Navy photograph.)

Los Angeles–class fast-attack submarine USS Helena (SSN-725), homeported at Naval Station Norfolk, is shown at the shipyard. It was visiting the shipyard to receive a series of technology upgrades. Helena is the fourth ship to be named after the capital of Montana. (US Navy photograph.)

This photograph depicts *Ohio*-class ballistic missile submarine USS *Maryland* (SSBN-740) at the shipyard after completion of its dry dock period in 2012. Normally, the *Maryland* has a Blue crew and a Gold crew. (Ballistic missile submarines have two crews to enable them to be underway indefinitely and complete their mission of strategic deterrence.) However, the two crews were combined for this extended maintenance period. (US Navy photograph by Shayne Hensley, NNSY photographer.)

Pictured is Norfolk Naval Shipyard's 250th anniversary logo. The shipyard celebrated this occasion throughout 2017.

Discover Thousands of Local History Books Featuring Millions of Vintage Images

Arcadia Publishing, the leading local history publisher in the United States, is committed to making history accessible and meaningful through publishing books that celebrate and preserve the heritage of America's people and places.

Find more books like this at
www.arcadiapublishing.com

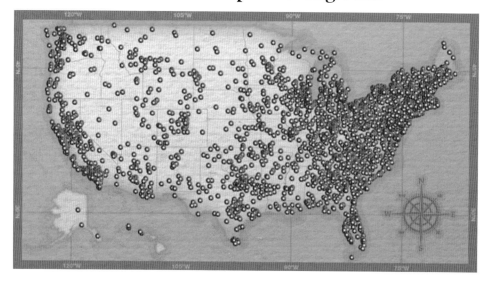

Search for your hometown history, your old stomping grounds, and even your favorite sports team.

Consistent with our mission to preserve history on a local level, this book was printed in South Carolina on American-made paper and manufactured entirely in the United States. Products carrying the accredited Forest Stewardship Council (FSC) label are printed on 100 percent FSC-certified paper.